MW01165103

# Do Animals Have Feelings Too?

By David L. Rice
Illustrated by Trudy Calvert

Dawn Publications

## Dedications

**To Venita, the most compassionate person I have ever known. —DR**

**I humbly dedicate this book with thanks to each person who encouraged me to pursue my "arts" desire, with blessings to Earth's creatures great and small, and with praise to God who made us all. —TC**



A Sharing Nature With Children Book



Library of Congress Cataloging-in-Publication Data

Rice, David L., 1939-
Do animals have feelings too?/ by David L. Rice ; illustrated by Trudy Calvert.—1st ed.
p.cm. – (A Sharing nature with children book)
Summary: Through facts and anecdotes, investigates the question of whether animals experience feelings such as compassion, loyalty, grief, joy, vengefulness, and helpfulness..
ISBN: 1-58469-003-8 (case)
ISBN: 1-58469-004-6 (paper)
1. Emotions in animals Juvenile literature. 2 Animal behavior Juvenile literature. [1. Emotions in animals. 2. Animals—Habits and behavior.] I Calvert, Trudy, ill. II Title. III. Series.
QL785.27.R53 2000
591.5—dc21 99-32162 CIP

DAWN Publications
P.O. Box 2010
Nevada City, CA 95959
800-545-7475
Email: nature@DawnPub.com
Website: www.DawnPub.com

Printed in China

10 9 8 7 6 5 4 3
First Edition

Design & computer production by Andrea Miles

Do animals have feelings? Until recently most scientists didn't think so. They thought just about all animals behaved instinctively—that most animals didn't feel happiness, or sadness, or enjoy life.

Many scientists are now changing their minds.

One of the best ways to learn about animals is simply to watch them closely over a long period of time. Here are some true stories that might show that animals have feelings similar to human feelings. Or, perhaps they are just examples of instinctive behavior. What do you think?

# Compassion

*Compassion is a feeling of sympathy for others who are feeling sad or are in some kind of trouble.*

- A visitor to Africa watched in horror as a Nile crocodile clamped its powerful jaws on the back legs of a small antelope and started dragging it into the river. Suddenly, a hippopotamus charged the crocodile. The crocodile released the antelope and quickly swam to safety. The hippo gently helped the injured antelope up the bank of the river, then comforted and protected it until the antelope finally died from its wounds.
- A man who raised musk oxen was feeding them when a pack of wild dogs charged toward him. The oxen formed a circle around the man to keep the dogs from getting to him. Oxen protect their young from attack in this same way.
- An injured bird fell into a chimpanzee pen in a zoo in Switzerland. Zoo workers expected the chimpanzees to eat the bird, since they often eat small animals. Instead, after picking up the bird, they handed it gently from one chimpanzee to another, and then gave it to a zoo worker.
- Another chimpanzee jumped a fence and waded out into a river to save a baby chimp who had fallen in and was drowning. The chimps did not know each other.
- A Texas game warden was surprised when he saw a blue jay giving food to another adult blue jay—something jays never do. He looked closer and saw that the jay receiving the food had a broken beak and couldn't get food on its own.
- A young couple was driving through a wild part of Africa when their car broke down. Getting out of the car, they heard a whimpering sound coming from the other side of a small hill. They discovered a small antelope that was leading a much larger wildebeest toward a waterhole. The wildebeest's eyes were swollen shut from a snake bite and it was in a great deal of pain. Whenever the bigger animal turned the wrong way, the antelope gently touched its neck to guide it.

*What is your experience? Have you seen animals behaving compassionately toward humans or other animals?*

*How about you? Tell about a time you felt compassion for a person or animal. What did you do to show your compassion? How did you feel after you helped them?*

A young Chicago couple decided to take their three-year-old son to the Brookfield Zoo for a family outing. When his parents were not looking, the active youngster climbed over a railing and fell two stories down into the gorilla pen, hitting his head. As the helpless child lay injured and unconscious, a young gorilla mother named Binti scooped up the hurt boy, and cradled him in her arms. She then carried him to the door of the pen and gently laid him where the zookeepers could rescue him. He was taken, still unconscious, to a hospital where after four days he completely recovered. If it had not been for Binti, he might not have been so lucky.

Thousands of people have visited the Brookfield Zoo to see the "compassionate gorilla."

A young male Canada goose, flying south with his flock for the winter, was shot in the wing by a hunter. He fell to the ground and into some bushes. He was nearly dead when a farmer found him several days later and nursed him back to health. The farmer had some tame female geese. The gander wasn't interested in them, but was attracted to a female at a nearby farm. The farmer bought the neighbor goose and the happy pair was seldom seen apart.

Next fall, flocks of wild geese again passed overhead, honking loudly on their way south. Unable to resist the call of the wild, the gander rose up and joined one of the flocks, leaving his tame mate behind. She was so upset that she refused to eat or go to the lake with the other geese.

Three days later, the gander was back. His loyalty to his mate was stronger than his instinct to fly south. The two geese spent the year together and raised chicks. Then one day the following fall, the female was accidentally killed. For hours the wild gander sat with his long neck resting quietly across her body. The next day a flock of wild geese flew overhead on their way south. The gander rose to join them. He never returned.

# Loyalty

*To be loyal means to be a friend to someone even when it would be easier not to be.*

- A group of false killer whales was in shallow water. One of them became seriously injured and could not swim out to sea. The other whales refused to leave the area and formed a protective circle around it. The U.S. Coast Guard, fearing that all the whales would become stranded when the tide went out, tried to herd the whales into deeper water, but they refused to leave their dying companion. Only when their friend was dead did they head for the safety of the open sea.

- A pair of swans, on their way south for the winter, stopped for the night on the Detroit River. During the night the water froze, and the female was stuck in the ice. Although the male was able to free himself, he would not leave. After a few days the female was rescued. She was sick and weak. It took a month for her rescuers to nurse her back to health. The male stayed nearby in the icy river, waiting for his mate. When the female was finally well enough to fly, the pair greeted each other happily and flew south. Swans are one of a number of birds that mate for life.

- Tarsiers, small monkeys who have large eyes and long tails, also mate for life. When a Tarsier is captured, its mate allows itself to be captured also.

- When a bottlenose whale is injured, the other whales in its pod will not leave until the injured whale either dies or is well enough to come with them. Before whale hunting was outlawed, whale hunters sometimes wounded a bottlenose whale and left it in the water. When its companions came to help, the whalers killed them all.

*What is your experience? Have you seen animals showing loyalty to other animals or humans? How about you? Tell about a time you felt loyalty for a friend or member of your family. How did it feel to be a loyal friend?*

# Grief

*To grieve means to feel sad and upset when you lose someone or something you love.*

- When an eight year old chimpanzee's mother died, he spent many hours sitting by her body. He showed no interest in eating or playing. Three days later, he climbed a tree and sat looking at the place where he and his mother had slept. A month later, he died, possibly of a broken heart.
- Every flock of crows has a leader. Being a boss crow is an important position, like being the mayor of a city. When a boss crow's mate died, he lost interest in being the leader. He perched apart from the others, and sat for days without eating. When a younger crow challenged him to become boss, the grieving crow gave up his leadership without a fight.
- Two dolphins were friends who had been together for several years. They often touched each other with their fins while swimming. When one of them suddenly died, the other refused to eat, swimming around and around with his eyes tightly shut.
- When Rix, a male chimpanzee, died from a fall into a deep gully, the members of his group became very upset. They screamed, hugged each other, and threw rocks. Then they spent several hours sadly looking at the body of their dead companion.
- In Italy, an elephant named Sandra and her trainer had been together for many years. Then the trainer got married and spent less time with Sandra. At that, Sandra refused to eat, became very depressed and soon died.

- Two cows had been together since they were babies. They were hardly ever seen apart. They even raised their calves together. Then one of the cows was sold and taken away. Her companion stood bellowing outside the barn door for weeks, with tears in her eyes, waiting for her friend to appear.

***What is your experience?*** *Have you seen animals that seemed to be grieving? How did they show their grief?*

***How about you?*** *Tell about a time when you lost someone or something you loved and felt grief.*

When a member of an elephant herd dies, the herd sometimes stops and forms a circle around it. They stroke their dead companion with their trunks, then tear off branches or pull up clumps of grass, which they lay around the body. They have also been observed forming a circle, facing away, as if they can't bear to look at their lifeless friend.

Elephants are able to recognize the bones of relatives. They smell them, turn them over and over and feel them with their trunks. One herd came upon the jawbone of a family member. After examining it at length, as if remembering times past, they moved on. But the dead elephant's young calf stayed behind for several hours, touching and moving the jaw of his mother. Another young elephant visited the skull of her mother every time she was in the area, much like a human visiting the grave of a loved one.

# Deceitfulness

*To be deceitful means to try to trick someone into thinking something is true when it is not.*

- When baboons want to warn their companions of a problem, they quickly turn their heads and stare towards the danger. One day a young baboon was attacked by an older, larger member of its group. The youngster's screams brought several adult baboons running to help. The attacker, not wanting to get beaten up, suddenly turned its head and stared into the distance at an imaginary danger. The other baboons, thinking there was danger, stopped and peered into the distance, ready to fight or run away. While they were trying to locate the danger, the troublemaker ran off, escaping punishment.

- A young beaver showed up late for a meal. There wasn't room for him to get to the food. Suddenly, he hit his flat tail loudly on the ground—a beaver's warning that danger is near. Immediately his whole family dived into the nearby pond for safety. The youngster walked over and ate.

- Many birds that build nests on the ground, such as quail and pheasant, trick their enemies to save their young. If a fox comes near, the mother bird walks away from the nest flapping one wing, pretending to be injured. The fox follows, thinking he'll get an easy meal. When mother quail has led the fox far enough away, she spreads her wings and flies.

- A zookeeper caring for gorillas noticed that one had its arm stuck in the bars of its cage, and was frantically trying to get free. The keeper rushed around the side of the building to get into the cage so he could help. Meanwhile, the gorilla easily freed its arm and hid by the entry door. As soon as the man entered the cage, the gorilla jumped out and grabbed him from behind. It was the gorilla's idea of a joke.

***What do you think? Which of these animals knew what they were doing and which were probably acting from instinct?***

***What's your experience? Has an animal ever tried to fool you, or have you seen one try to fool another animal?***

Prairie dogs live in "towns" and are always on the alert for danger. When they spy a coyote, they call a loud warning and dive into their burrows for safety. Sometimes two coyotes team up to outsmart the prairie dogs. One coyote hides while the other approaches the prairie dog town, causing the prairie dogs to dive underground. While the prairie dogs are underground, the hidden coyote moves much closer and hides again. Meanwhile, the other coyote walks past all the burrow entrances and, apparently giving up, keeps going. When it is far away, the prairie dogs mistakenly think that the danger is over. One of them gives the "all clear" signal and they scamper from their burrows. The hidden coyote jumps out and catches dinner for the team of cunning coyotes.

American Eagle parents take turns sitting on their eggs for 35 days before they hatch. Eaglets are very small and require a great deal of care and training. At first, the parents put food right into the chicks' mouths. When they are a little older the parents bring fish broken into small pieces, which the chicks must eat on their own. Finally the eagles bring whole fish and teach their young how to tear it into small pieces for themselves.

They teach the growing eaglets to build wing muscles by jumping up and down and flapping their wings every day. They also require the eaglets to grab sticks with their beaks and talons and hold on tightly. When it is time for young birds to take their first flight, the parents get them to jump out of the nest by holding pieces of food just beyond their reach. The parents reward them with pieces of food after they return from their first flight.

# Parental Devotion

*Parental devotion is the love and care a parent gives to its young—feeding, protecting, comforting and teaching them.*

- Jewelfish are one of only a few kinds of fish that care for their offspring. Babies follow their mama much like chicks follow a hen. At dark, the father makes a nest in the sand for the little jewels. The mother calls the young fish by flashing her colored fins. If one or two don't come swimming, father searches out the stragglers, sucks them into his mouth and spits them into bed.
- A mother stork was raising babies in a nest on top of a building in Denmark. The building caught on fire, but she refused to abandon her chicks. As the fire came closer, she covered her young to protect them and wildly beat her wings to keep the smoke away. When the fire was finally put out, mother stork was black from smoke and soot, but her babies were safe.
- In Uganda, a young elephant wandered too close to a steep river bank. The bank suddenly gave way under the elephant's weight, and the youngster fell into the swift river below. After being rescued by a couple of "aunts," his mother came running. She held him close to her, made worried mother sounds, and carefully felt all over his body with her trunk. When she was sure he was unhurt, mother elephant gave him a hard swat with her trunk and chased him away from the river.
- Mountain goats live on steep, rocky mountains. Mother goats watch their youngsters very carefully, and stay on the downhill side of their kids to stop them if they fall. When a young goat takes a hard fall, the mother cries out loudly, rushes to her youngster, gently licks it, and allows it to nurse.
- If attacked, herds of giraffes can run very fast—up to 35 miles an hour—on their six-foot long legs. However, when a lion attacked a herd that included a baby, the young giraffe could not keep up with the adults. Its mother tried to push it with her nose, but it just couldn't run any faster. So mother stopped and bravely faced the lion. Every time the lion tried to get the young giraffe, the mother kicked with her front feet. After about an hour the lion gave up and decided to find supper elsewhere.

***What do you think? Were these animal parents feeling the same kind of love and devotion that human moms and dads feel for their children?***

***What is your experience? Have you observed animal parents showing love and devotion to their babies?***

# Romantic Love

*Romantic love is the affection that males and females share.*

- When a male raven finds a female to his liking, he prepares bits of food as gifts ahead of time, and hides them. When she comes by, he grabs the food, marches up to her and offers it, making sounds to encourage her to take it. If the female is interested she lowers her body, flaps her wings and takes the food. Ravens go through an engagement period of about a year. While engaged, they hunt for food together, practice nest building, fight off attackers and groom each other. During this period ravens may change partners in an attempt to find a more suitable mate. But once the engagement period is over, they are a pair for life—and ravens can live 50 years or longer.

- A scientist wanted to find a mate for his caged male cockatoo. A young female cockatoo with beautiful feathers was brought to him. The male refused to have anything to do with her. He even pretended she wasn't in the cage. Some weeks later the scientist obtained an older female. She had many feathers missing and there were wrinkles around her beak. She was put into the cage with the male. He was immediately interested and acted as if he had found the love of his life. They raised many young cockatoos together.

- A pair of wolverines was seen rolling on the ground and playing together. The female would tease the male and run off, daring him to chase her. When he didn't follow, she came back and bumped him with her hip. Even though wolverines are one of the fiercest animals alive, they have their romantic side—but only for a few days a year.

***What do you think? Is the romantic behavior of these animals similar to that of humans?***

Emperor penguins spend much of each year apart from their mates. When they return to their winter home in the Antarctic, their first job is to find each other amongst thousands of other penguins whom, at least to humans, look very much alike. They find each other by singing. Each couple has its own song. When one of them hears its mate's song, it joins in and the two rush toward each other. They stand in front of each other, bow their heads and lean against each other. They often remain quietly leaning against each other for hours.

In the forests of Canada, a Cree Indian made his living trapping foxes, martin and ermine. He stored the animal pelts in his cabin. One day a wolverine ate most of the animals caught in the traps, ruining their fur. The trapper knew the old saying: "If a wolverine moves in, either it must be killed or the trapper must leave." He was determined to kill the wolverine.

The next morning he set out with his large wolfhound. Soon his dog growled and ran into the forest. Suddenly there was a wild commotion—the dog and the wolverine were fighting. The dog was badly hurt and soon died. The wolverine ran off.

More determined than ever, the hunter built wolverine traps until late afternoon. He decided to camp under a large tree. He hung his snowshoes in the tree and kept his gun close. The next morning, he discovered that his snowshoes had been completely destroyed by the wolverine. Unable to return without snowshoes, he hid his gun and equipment and went looking for willow branches to make new ones. When he returned, he found that the wolverine had destroyed his blanket and carried off his matches and gun. When he finally arrived at his cabin, he discovered that the wolverine had broken in, ruined all of the animal pelts, destroyed most of his possessions, and had eaten or carried off all of the food.

# Vengefulness

*To be vengeful is to try to get back at someone who has hurt you.*

- A trainer at an oceanarium had been teasing Ola, a large Orca whale. He knew that she didn't like it, but he continued. One day while he was teasing her, Ola had had enough. She pushed the trainer to the bottom of the tank and held him there. It took the other trainers over five minutes to get Ola to let him go. Fortunately he was wearing diving gear, or he would have drowned. Ola has never done anything to hurt anyone else.

- In a wild animal preserve, a giraffe was standing in the middle of the road. When the driver of a car honked loudly at it, the giraffe knocked the car over and started kicking it. On another occasion, a car honked and flashed it lights at two giraffes in the road. One got off the road. The other kicked and smashed the car's radiator.

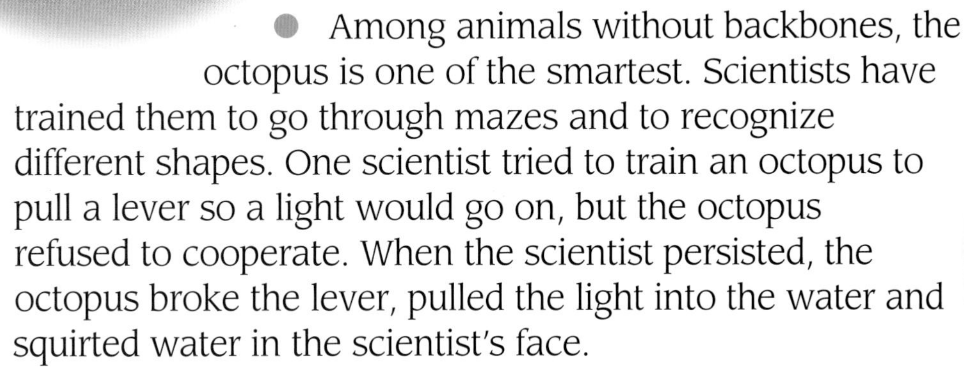

- A pair of swallows had just finished building their nest. A sparrow flew into it. The swallows tried to make the sparrow leave but the bully refused, intending to take over the nest. The swallows finally flew off, but soon returned, followed by a whole flock of swallows. Each swallow put a mouthful of mud in the hole, sealing the entrance to the nest. That nest became a tomb for the sparrow.

- Among animals without backbones, the octopus is one of the smartest. Scientists have trained them to go through mazes and to recognize different shapes. One scientist tried to train an octopus to pull a lever so a light would go on, but the octopus refused to cooperate. When the scientist persisted, the octopus broke the lever, pulled the light into the water and squirted water in the scientist's face.

*What is your experience? Have you seen animals acting in a vengeful way?*

*What do you think? Why do humans hire police officers and others to help them with problems, rather than getting revenge themselves?*

# Joy

*Joy is a feeling of great happiness.*

- Otters love to play, wrestle, chase each other and pretend to fight. They play with any playful animal—not just other otters. A favorite game is water sliding. Father usually starts the game, tucking his legs in and diving head first down a steep muddy bank, hitting the water with a loud splash. Mother and then the young otters follow. Many otter families will keep at this game for hours. Otters are usually friendly—but if necessary, they prove themselves to be among the most fearless fighters in the animal kingdom.
- Pandas and brown bears frequently roll down hills, apparently just for the fun of it.
- Gorillas swing their children around and wrestle with them much the way humans do.
- Ravens have been seen rolling over on their backs and sledding down snowy hills.
- Sea lion mothers don't seem to mind if their children play with their food. They often toss their fish into the air and catch it with their flippers several times before eating it.
- A hummingbird was observed playing in a stream of water coming from a hose. It would land on the water close to the hose and let itself be carried along. Just before hitting the ground, it would fly back to the hose for another ride.
- A travelling circus camped next to a playground. A young elephant watched children swinging on the playground swings. The elephant was not tied up. It went to the playground and tried to sit in one of the swings. Unfortunately the swing was not elephant-sized. After several unsuccessful tries the disappointed youngster gave up.
- When a strong wind blows, right whales sometimes go for a "sail." They raise their tails straight up and let the wind push them along like a sailboat. When they get too close to shore, they swim out and sail again. They play like this for hours without tiring of the game.

***What is your experience? Have you seen animals that seemed to be joyful?***

***How about you? Tell about a time you felt a great deal of joy and happiness.***

Alaskan bison love to skate across frozen lakes. Taking turns, they run down a hill and onto the ice. With their legs held stiff and their tails sailing through the air, they spin around and around. When they finally come to a stop, they let out loud bellows! Then, like eager children, they make their way back to shore and up the hill for another turn.

# Helpfulness

*To be helpful is to assist someone, especially with something they could not do alone.*

- An animal scientist was surprised when he came upon two rats, because one was holding the other's tail as they walked along. The scientist discovered that the second rat was blind and was being guided by the first rat. Other observers have reported seeing mice, sheep, and even fish leading their blind companions to food and water.

- Scientists who study bats have noticed that they often share food. But they don't share with just anyone. Rather, they share with their parents, siblings and other relatives. They also share with their neighbors—bats that they roost next to, or who have given them food in the past.
- Mexican jays live in flocks of up to twenty birds. They often fight and quarrel. But when young jays begin to hatch, the squabbling stops and the group works together to raise the young. Birds that don't have their own babies help to feed the babies of their neighbors.
- When a raven finds a dead animal, it flies back to its flock, loudly giving raven yells. This tells the others that food has been found. They follow their generous neighbor back to dinner.
- African spiny mice help each other give birth. Several females stay with the mother when she goes into labor. They help the mother give birth and clean up the newborn babies. Dolphins, elephants and llamas also help other mothers give birth.
- In herds of elephants, grandmothers and aunts often help take care of the youngsters.

***What do you think?*** *Which of these animals helped because they were caring and which probably acted out of instinct?*

***How about you?*** *Tell about a time you were helpful to an animal or person who needed help.*

When a baby gnu is born it cannot stand up for almost 15 minutes. During this time a hyena, wild dog or lion could easily kill it. On one occasion a baby had just been born when a large lioness crept toward the helpless baby. A gnu herd will usually run off in fear at the sight of a hungry lion. However, instead of running, some of the gnus lined up to form a wall between the lioness and the baby. The lioness growled fiercely as she inched closer, but the gnus would not budge. Finally she turned and walked away—it wasn't worth the fight.

# Democratic Choices

*In a democracy people make decisions together, and elect their leaders.*

- The boss of a group of Japanese monkeys is usually the one that is strongest and the best fighter. But sometimes it's not that simple. In one group of about 300 monkeys, the old leader was injured and defeated by a young male. The monkeys didn't like the new leader, however, and the group refused to follow him. The young male tried beating the others into obeying him, but they would not follow him. This group, at least, showed that it had some rights in selecting its leader.

- When crows fly for long distances, each crow makes one of two different calls. One call says, "I'm tired and want to stop and rest." The other call says, "I want to keep flying." When about three fourths of the crows call out to stop, the whole flock lands and rests.

- When a swarm of bees separates from its old hive and leaves with a new queen, thousands of bees must find a good place to start their new hive or the whole swarm could die. Somehow, about 40 bees are selected as scouts and are sent out to find a good hive spot. As the scouts return, they tell the other scouts about the place they found. Each scout tries to convince the other scouts that theirs is the best spot. Small groups form and each group tries to convince the others. Meanwhile the whole swarm waits, getting hungrier. Usually it takes a day or two before all 40 scouts agree on the new location. Then the swarm moves into its new home.

- South American army ants travel in "armies" of up to twenty million ants. When their number becomes too large, they divide in two groups. Shortly before this happens, several new queens hatch. Only one of them can become queen of the new group. Instead of fighting until one kills the other, however, each queen tries to convince her fellow ants to follow her. The one who attracts the most followers becomes the new queen.

*What do you think? Which of these animals knew they were making a choice and which were probably acting from instinct?*

*What is your experience? Have you seen animals act in a democratic way?*

Prairie dogs live in "towns," and they have a "mayor" that they "elect." If a young male prairie dog wants to become a leader, he has to convince other prairie dogs to let him take over. He can't just defeat the old leader in a fight, because even if he should win, the other prairie dogs may choose to follow the old mayor to another area and start a new town.

So, the ambitious young prairie dog needs to campaign for himself. He does this by being friendly to a few prairie dogs at a time, trying to get them to like him better than the old mayor. Finally, when he thinks that most of the other prairie dogs will favor him, he challenges the current leader, who must then either fight or leave town in the hope that the others will follow.

# Cleverness

*Cleverness is being smart, particularly being able to figure out how to do something in a new way.*

- A crow watched while two men fished through a hole in the ice. The men had rigged their fishing line so that when a fish pulled on the line a red flag would go up. This allowed them to wait in their warm car until they saw the red flag. Then they would get the fish and put more bait on the line. After watching for a while, the crow evidently decided that it was her turn. The next time the flag went up, the crow went over to the hole, grabbed the line with its beak and backed away. The astonished fishermen watched as the crow stood on the line while it walked back to the hole and grabbed the line again. It repeated this routine until the fish flopped up on the ice. The fishermen were speechless as they watched the crow have a fish for dinner.

- Ravens, which are large members of the crow family, have been seen leading a wolf or coyote to a dead animal whose tough hide the raven could not peck through. After the coyote or wolf tears open the animal and eats, the raven takes its turn. Some hunters claim that ravens have led them to deer and moose, presumably with the idea that the birds would share in the kill.

- An oceanarium in San Francisco had a problem with people throwing trash into the dolphin pools. One of the trainers came up with an idea to solve the problem. The dolphins were trained to bring the trash to the trainers, and were given a piece of fish for each piece of trash. The plan worked marvelously. The dolphins loved the game. But when one clever dolphin found a large brown paper bag on the bottom of the tank, it wasn't willing to settle for just one piece of fish. By tearing off one little piece one at a time this genius was able to get several pieces of fish before the trainers caught on!

- Pelicans have discovered an easy way to catch fish. A number of pelicans get together and swim in a half circle toward the shore making as much commotion as they can. This frightens the fish and herds them toward the beach. When the pelicans get to shallow water, they wade shoulder to shoulder, flapping their wings. The trapped fish are then easily scooped up in their large beaks.

*What do you think? Were these animals thinking in the same way humans think?*
*What is your experience? Have you seen animals behave in an especially clever way?*

A rhesus monkey escaped from the monkey and ape pen at the Bronx Zoo one night. It was several days before he was captured in a nearby park. The zookeepers checked the pen carefully but were unable to discover how he had escaped. Confident that it could not happen again, they released the rhesus back into the pen. The next morning he was gone again! The escape artist was finally captured again, and again zookeepers checked the ape enclosure for possible escape routes. When they couldn't find any, one of the workers stayed up all night to watch him. It was a boring night watching the monkey sleep. Then, at dawn, the monkey took a banana that he had hidden the day before and went to the edge of the water-filled moat that separates the apes and monkeys from the moose pasture. He waved the banana back and forth in the direction of a large moose on the other side of the moat. Soon the moose came swimming over. After giving it the banana, the monkey jumped on the moose's back and got a ride to the other side of the moat where he could easily escape.

Two chimpanzees named Sherman and Austin were taught the meaning of over a hundred symbols that stood for English words. Then they were taught to use computer keyboards that contained these symbols. Scientists wanted to find out if the two chimps were really able to communicate with each other using the symbols on the keyboard. They put a keyboard and computer screen in both chimps' rooms, which were next to each other. While Austin was away, the scientists let Sherman watch them put food in a box and lock it with a key. They put the locked box in Sherman's cage, but they put the key in Austin's toolbox. When Austin returned, Sherman typed out the symbols for "PLEASE AUSTIN GIVE KEY." Austin read the message, found the key and gave it to Sherman. Sherman opened the box and shared the food with Austin.

# Communication

*To communicate is to let someone else know something, often through writing, speaking, sounds, facial expressions and body movement.*

- Koko, a young gorilla, was taught over 600 words in sign language. One day when her teacher was showing a visitor how smart Koko was, the gorilla would not cooperate. When told to touch one part of her body, she would touch the wrong part.

  "Bad gorilla!" Her teacher signed to her.

  "Funny gorilla." Koko signed back, smiling.

  When one of her teachers acts silly, Koko signs, "That funny."

- A chimpanzee named Lana also types words using a computer keyboard. She has even taught herself to read the sentences she writes. If she makes a mistake while writing a sentence she erases it and starts again. One time she wanted an orange that her teacher was eating. She knew the word for apple and that orange was a color, but had not been taught that orange was also the name of a fruit. When they showed her an orange, she typed, "Please give apple that is orange." Another chimpanzee named Washoe makes up words of her own. She called a watermelon a "drink fruit," and a swan a "water bird." A monkey that she didn't like she called "dirty monkey."

- An African gray parrot at the University of Arizona has learned to say over 100 words. He can count to 6 and can say which objects are larger or smaller and what color they are. If he bites someone, his teacher says, "Bad boy!" and starts to leave. The parrot will reply, "Come here. I'm sorry."

- Dolphins communicate with each other using at least 18 different sounds. In one test, a dolphin used whistles and clicks to tell another dolphin to push a lever so they could get food.

- Crows use as many as 50 different calls to communicate with each other. Some gulls use over 60 different calls.

*What do you think? Do these animals communicate the same way humans communicate?*

*What is your experience? Have you seen animals communicating with each other?*

# Love of Beauty

*To love beauty is to have strong feelings of peace and joy from something that is beautiful, such as nature, art or music.*

- In many bird species, the males attract females with their brightly colored feathers. Male bowerbirds, however, do not have colorful feathers, but instead decorate special areas, called bowers, under trees and bushes. They use flowers, seeds, berries, feathers, shells and any colorful object they can find to make their bower beautiful. Some bower birds crush berries to make paint, pull bark from trees to make paint brushes, and paint their bowers. They often step back to observe their work, like an artist who is not quite satisfied, and then rearrange some part of their display. The female bowerbird seems to choose her mate on the basis of how attractive she finds the bower.
- Many apes and monkeys seem to enjoy drawing. One chimp would beg visitors for a pencil and paper, even preferring it to food. She would sit in the corner and draw. Another chimp would go into a rage if he was interrupted before he was finished.
- A man named Gerald Durrell had a pet pigeon that loved music. Most of the time it would listen quietly, leaning against the record player. When marches were played, however, it would loudly coo and stomp back and forth. If a waltz was being played, it would coo softly, bow and turn in time with the music.
- Humpback and right whales are the only known mammals, besides humans, that sing real songs. Some of their songs last for 30 minutes. All the whales in one group sing exactly the same song. They only sing for six months of each year. During that six months they add to and change their song. When they start singing again the next year they begin with exactly the same song as they left off with the year before. Some scientists wonder if these songs might tell a story.
- One gorilla who was learning sign language seemed to love the singing of Pavarotti, the famous opera singer. He even preferred to remain inside during a Pavarotti television performance rather than go outside to play. He also enjoys tapping on pipes and strumming the strings of burlap sacks.

*What do you think? Were these animals appreciating beauty and music or were there other reasons for their behavior?*

*How about you? Tell about some sights and sounds in nature that you especially enjoy.*

One evening just before sunset on the Gombe Animal Reserve in Africa, a chimpanzee was observed climbing a hill that overlooked a beautiful lake. Soon, another chimp climbed the hill. The two chimps greeted each other with quiet grunts. Taking each other's hand, they sat down and watched the sun set over the lake. Bears have also been seen sitting quietly, watching the sun go down.

## Resources

**Animals Can Be Almost Human** (The Reader's Digest Association, 1979) *loyal gander*

**Crows, Jays, Ravens and Their Relatives** by Sylvia B. Wilmore (Eriksson, 1977) *romantic ravens*

**Elephant Memories: Thirteen Years in the Life of an Elephant Family** by Cynthia Moss (William Morrow, 1988) *grieving elephants*

**Fish Facts and Bird Brains: Animal Intelligence** by Helen Roney Sattler (Lodestar Books, 1984) *caring jewel fish father; playful whales, clever crow, musical whales, clever dolphin, communicative chimpanzee, tricky coyotes, teasing gorilla*

**Friendly Beast, The** by Vitus B. Droscher (Dutton, 1970) *grieving boss crow, clever rhesus monkey, tricky beaver*

**Good Natured** by Frans de Waal (Harvard University Press, 1996) *loyal killer whales, tricky gorilla*

**Just Like An Animal** by Maurice Burton (Charles Scribner's Sons, 1978) *jilted elephant (in grieving), helpful rats, playful raven*

**Marvels and Mysteries of Our Animal World** (Reader's Digest Association, 1964) *caring mother stork, loyal bottlenose whales, loyal swan, worried mother elephant, vengeful wolverine, caring eagle parents*

**My Family and Other Animals** by Gerald Durrell (Penguin Books, 1977) *musical pigeon*

**Our Amazing World of Nature: Its Marvels and Mysteries** (The Reader's Digest Association, 1969) *compassionate blue jay, vengeful swallows*

**Real Animal Heroes** by Paul Drew Stevens (New American Library, 1988) *compassionate hippo, protective musk oxen*

**Through Our Eyes Only?** by Marian Stamp Dawkins (W.H. Freeman, 1993) *deceitful baboons, sharing vampire bats*

**When Elephants Weep** by Jeffrey Moussaieff Masson and Susan McCarthy (Delacorte Press, 1995) *compassionate chimpanzees, loyal dolphins, protective mother giraffe, grieving chimpanzee, romantic parrot, vengeful false killer whale, joyful bison, apologizing parrot, nature loving chimpanzees, Pavarotti loving parrot*

When David Rice was seven years old, he observed a small dog trying to wake its mother which had just been killed by a passing car. As he watched the grieving puppy's vain attempts, he was struck by the depth of its sadness and pain. David's lifelong interest in feelings—both animal and human—comes through in his other books, ***Lifetimes*** and ***Because Brian Hugged his Mother***, published by Dawn Publications. ***Lifetimes*** introduces some of nature's longest, shortest and most unusual lifetimes and the lessons we can learn from them. ***Because Brian Hugged His Mother*** shows how a chain reaction of kindness can spread through a whole school and community as a result of a single hug.

Trudy Calvert grew up in a rural, wooded area, near a creek where some of her first friends were tadpoles, minnows, ducks, squirrels and snakes. She raised a baby robin, and later a coyote pup that had been hit by a car. As an art teacher, and now as a full time professional artist, Trudy has a passionate love of nature and animals. She is past president of the Indiana Wildlife Artists, a group that promotes wildlife art. Often she can sense what animals are feeling and responds without needing spoken words.

## Also by David L. Rice

***Lifetimes*** introduces some of nature's longest, shortest, and most unusual lifetimes, and what they have to teach us. This book teaches, but it also goes right to the heart. (Teacher's Guide available.)

***Because Brian Hugged His Mother.*** Brian's hug set in motion a series of unselfish acts that reached more people—and even animals—than he could ever know.

## Other Distinctive Nature Awareness Books from Dawn Publications

***Stickeen: John Muir and the Brave Little Dog*** by John Muir as retold by Donnell Rubay. In this classic true story, the relationship between the great naturalist and a small dog is changed forever by their adventure on a glacier in Alaska.

***My Monarch Journal*** by Connie Muther, shows in stunning photography the metamorphosis of a tiny egg to a caterpillar, then to a chrysalis, and finally to a beautiful butterfly—one of nature's most astonishing miracles. It is a guide for students to follow the development of their own monarch—and in the process, to gain a profound appreciation for these remarkable tiny beings. Available in both student (32 p.) and parent-teacher (52 p.) editions.

***With Love, to Earth's Endangered Peoples*** by Virginia Kroll. All over the world, groups of people, like species of animals, are endangered. Often these people have a beautiful, meaningful relationship with the Earth, and with each other. This book portrays several of these groups of people, with love. (Teacher's Guide available.)

***A Tree in the Ancient Forest*** by Carol Reed-Jones, uses repetitive, cumulative verse to show graphically the remarkable web of interdependent plants and animals that all call a big old tree home.

***Animal Acrostics*** by David Hummon, is a unique celebration of animals through word play. In an acrostic poem, the first letters of the poem's lines, when combined, spell out the creature's name. Part riddle, these playful verses offer striking images of the world's creatures.

***In a Nutshell*** by Joseph Anthony. Striking illustrations and simple text follow the life cycle of an acorn that becomes a magnificent oak—and beyond. It gently suggests answers to the greatest of life's questions—Who are we? Why are we here? Where do we come from and where we go? Here, in a nutshell, is a tale about life.

**Dawn Publications is dedicated to inspiring in children a deeper understanding and appreciation for all life on Earth. To order, or for a copy of our catalog, please call 800-545-7475. Please also visit our web site at www.DawnPub.com.**